# SCIENCE FACTORY
# LIGHT & SIGHT

## JON RICHARDS

**PowerKiDS**
press™

New York

Published in 2008 by The Rosen Publishing Group, Inc.
29 East 21st Street, New York, NY 10010

Design:
David West Books

Designer:
Flick Killerby

Illustrators:
Ian Moores and Ian Thompson

Photographer:
Roger Vlitos

Library of Congress Cataloging-in-Publication Data

Richards, Jon, 1970–
Light & sight / Jon Richards.
p. cm. — (Science factory)
Includes index.
ISBN-13: 978-1-4042-3904-3 (library binding)
ISBN-10: 1-4042-3904-9 (library binding)
1. Light—Juvenile literature. 2. Light—Experiments—Juvenile literature. 3. Sunshine—Juvenile literature. I. Title. II. Title: Light and sight.
QC360.R52 2008
535.078—dc22
2007016691

Manufactured in the United States of America

# ABOUT THE BOOK

*Light and Sight* examines the basic aspects of light, as well as its more complex and practical uses. By following the projects carefully, the readers are able to develop their practical skills, while at the same time expanding their scientific knowledge. Other ideas then offer them the chance to explore each aspect further to build up a more comprehensive understanding of the subject.

# CONTENTS

### YOUR FACTORY
Hints and tips on how to prepare for
your experiments
pages 4-5

### LIGHT FOR LIFE
See how important the sun is for all life
pages 6-7

### IN CAMERA
Build a pinhole camera
and see the world
upside down
pages 8-9

### IN THE DARK
Explore the dark world of shadows
pages 10-11

### BOUNCING LIGHT RAYS
See over walls by making
light rays turn corners
pages 12-13

### TRANSPARENT, TRANSLUCENT, AND OPAQUE
See how light passes
through solid objects
pages 14-15

### BENDING LIGHT
Build a microscope and
see how bending light
can be useful
pages 16-17

### BRINGING IT CLOSER
See the world enlarged with
your own telescope
pages 18-19

### SPLITTING LIGHT
Split up light to make
a rainbow
pages 20-21

### MIXING LIGHT
Mix colors of light to
form multicolored
patterns
pages 22-23

### SEPARATING COLORS
Split up different-colored
inks using water
pages 24-25

### LIVING PICTURES
Create your own moving
characters
pages 26-27

### BEAMS OF LIGHT
Find out how light is used
to carry information along cables
pages 28-29

### GLOSSARY
pages 30-31

### INDEX
page 32

# YOUR FACTORY

BEFORE YOU START any of the projects, it is important that you learn a few simple rules about the care of your Science Factory.

● Always keep your hands and work surfaces clean. Dirt can damage results and ruin a project!

● Read all the instructions carefully before you start each project.

● Make sure you have all the equipment you need for the projects (see checklist opposite).

● If you don't have the right piece of equipment, then improvise. For example, a liquid detergent bottle will do just as well as a plastic drink bottle.

● Don't worry if you make mistakes. Just start again — patience is very important!

● Now you are ready to start. Turn the page and remember to have fun in your Science Factory!

Equipment checklist:
- ● Paper and cardboard
- ● Shallow trays
- ● Tissue, blotting, and tracing paper
- ● Scissors, adhesive tape, and glue
- ● Cotton balls and sprout seeds
- ● Paper fasteners and rubber bands
- ● Flashlights and rulers
- ● Cardboard boxes and tubes
- ● Modeling clay and beads
- ● Pencils, felt-tip pens, and paints
- ● Small mirrors and blankets
- ● Colored plastic wrap
- ● Plastic bottles and corks
- ● Eyepiece and convex lenses
- ● Drinking straws
- ● Large glass bowl
- ● Bulldog clips, pins, and nails
- ● Glass jar
- ● Milk

### WARNING:

Some of the experiments in this book need the help of an adult. Always ask a grownup to give you a hand when you are using electrical appliances or sharp objects, such as scissors.

# LIGHT FOR LIFE

**WHAT YOU NEED**
*Shallow tray
Cotton balls
Sprout seeds
Cardboard*

THE BRIGHTEST AND MOST OBVIOUS SOURCE OF LIGHT IS OUR NEAREST STAR, THE SUN. The sun supplies us, as well as plants and animals, with light and warmth — without it we would not be able to survive! This experiment offers an introduction to the world of light by showing you just how important sunlight is in keeping plants alive.

## GROWING PLANTS

**1** Spread the cotton balls in the tray. Moisten the cotton balls with water and scatter the sprout seeds over them. Leave it until the seeds have sprouted.

**2** Ask an adult to cut out your initials or a pattern from the cardboard. Place this over the seedlings.

**3** Leave the tray in a sunny spot for about two weeks, keeping the cotton balls moist with water.

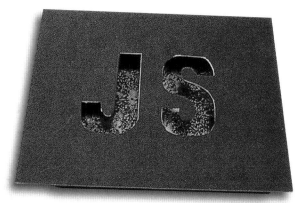

**4** When the sprouts have fully grown, remove the cardboard. You will see that the sprouts exposed to the light by your cutout initials are much greener.

## TURNING SEEDS

Plants always grow toward the light. You can see this by placing a seedling on a windowsill. After one week you will see that it has started to grow toward the light. Turn the pot around, and it will start to grow back toward the light.

## WHY IT WORKS

Plants use sunlight, a gas in the air called carbon dioxide, and water to make food that they use to grow. At the same time, they release a gas called oxygen into the air. If the plants are kept in the dark, they cannot make their food, so they will wither and be paler than those exposed to the sun.

SUNLIGHT

CARBON DIOXIDE ABSORBED

OXYGEN RELEASED

WATER ABSORBED

7

# IN CAMERA

## WHAT YOU NEED
*Cardboard box*
*Sharp pencil*
*Tracing paper*
*Paints*
*Adhesive tape*
*Blanket*

LIGHT FROM THE SUN, AS WELL AS ALL OTHER FORMS OF LIGHT, travels in straight lines as light rays. As a result, light cannot naturally turn corners (if you turn to pages 12-13, you will see how light can be made to go around objects). This creates events such as shadows (see pages 10-11). It can even appear to turn the world upside down, as this project to make your own pinhole camera reveals.

## PINHOLE CAMERA

**1** *Ask an adult to cut the back off a cardboard box. Decorate the box using the paints.*

**2** *Ask an adult to make a small round hole in the front of the box using a sharp pencil.*

## BLURRED IMAGE

Using the sharp pencil, ask an adult to make the hole in your camera slightly larger. The image on the screen will become fuzzy. This is because the larger hole lets more light rays enter the camera. These light rays then hit the screen at lots of different angles, making the picture become fuzzy.

**3** Tape a sheet of tracing paper to the back of the box.

**4** Pull the blanket over your head and the back of the camera. Point the camera at a bright window and you will be able to see an upside-down image of the window on the tracing paper in the back of the box.

## WHY IT WORKS

The opening of the pinhole camera is very small. When rays of light from the window enter the camera, they cross over because they travel in a straight line. As a result, the image on the screen is upside down. The same thing happens with your eye. Light rays cross over as they pass through the pupil and enter the eyeball. When they hit the retina, they form an upside-down image. This image is sent to the brain, which turns the picture around again, letting you see the world the right way up!

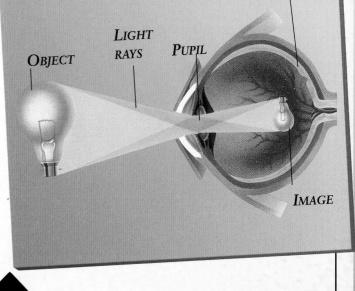

RETINA

LIGHT RAYS

PUPIL

OBJECT

IMAGE

# IN THE DARK

NOW THAT YOU KNOW THAT LIGHT RAYS TRAVEL IN STRAIGHT LINES (see pages 8-9), you can explore some of the effects of this. One of these effects is shadows. Look at your feet, and if it's a sunny day or you're in a bright room, you will see your shadow stretching off in the opposite direction from the light. This shadow is formed because your body blocks light rays and stops them from lighting up the dark area. Make your own puppet theater and have some fun experimenting with shadows.

## WHAT YOU NEED

*Dark cardboard
Paper fasteners
Tracing paper
Glue
Drinking straws
Flashlight*

## SHADOW PUPPETS

**1** *Ask an adult to cut out the body parts of the puppets from the dark cardboard, as shown above.*

**2** *Join the body parts together using the paper fasteners, making sure that the limbs can be moved.*

**3** *Glue the straws to the feet of the puppet. These will help you to control the puppet.*

## BLURRING SHADOWS

*Move the puppets away from the screen, and their shadows become blurred, with an area of half-shadow around the edge. This half-shadow is called a penumbra.*

**4** Cut a semicircle out of a sheet of cardboard. Tape tracing paper over this hole to make the screen. Tape some cardboard supports to the back to keep the screen upright.

**5** Shine the flashlight from behind the screen and entertain your friends by having your shadow puppets perform a story for them.

## WHY IT WORKS

The dark area, or shadow, is caused by an absence of light. Rays of light from your flashlight are blocked by your puppet before they can reach the screen. This creates an area of the screen in the shape of your puppet that is darker than the rest of the screen. The completely dark part of the shadow is called the umbra.

UMBRA

# BOUNCING LIGHT RAYS

THE PREVIOUS EXPERIMENTS HAVE SHOWN HOW LIGHT RAYS TRAVEL IN STRAIGHT LINES. However, with the right equipment, light rays can be made to go around corners! Shiny surfaces, such as mirrors, bounce light rays off them. This is called reflection. Use mirrors to build a periscope and see how you can make light turn corners.

**WHAT YOU NEED**
*Cardboard*
*Ruler*
*Pencil*
*Two small mirrors*
*Adhesive tape*

## UP PERISCOPE

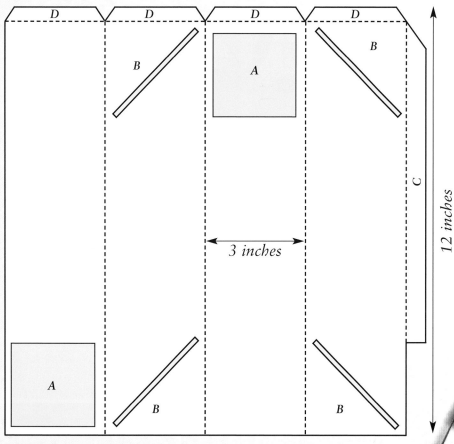

**2** *Tape the edges of the mirrors. The mirrors should be slightly wider than your periscope.*

**3** *Fold the periscope and tape it together to hold tab C in place. Tape some cardboard on top of the periscope using the four D tabs.*

**1** *Using a ruler and pencil, draw an open plan of your periscope — like the one shown here — on the cardboard. Ask an adult to cut out the plan and two windows (A) and four slits to hold the mirrors (B).*

## WHY IT WORKS

The shiny surfaces of the mirrors in your periscope reflect light rays. As a result, the rays of light that would normally travel over your head bounce off the mirrors, down through the periscope, and into your eye. This lets you see over any obstruction.

LIGHT FROM OBJECT

MIRROR

LIGHT REFLECTED OFF MIRROR INTO EYE

**4** Slide the mirrors into the slots and use some adhesive tape to hold them in place. Now look through the lower mirror, and you will be able to use your periscope to see over tall objects such as a fence or a wall.

## ODD REFLECTIONS

Look at your reflection on both sides of a metal spoon. Because the surfaces are curved, they will reflect light rays in different directions, giving you what looks like a funny-shaped face — one of the sides will even turn the image of your face upside down!

# TRANSPARENT, TRANSLUCENT, AND OPAQUE

**WHAT YOU NEED**
*Glass jar*
*Water*
*Milk*
*Flashlight*

YOU'VE ALREADY SEEN HOW LIGHT IS BLOCKED BY SOME OBJECTS and reflected by others. Now you can see how light can actually pass through things! When light can pass through something, that object is called see-through, or transparent. However, some objects appear cloudy and only let some light through. They are called translucent. Other objects that let no light through are called opaque. This project will help you explore things that are transparent, translucent, and opaque.

## CLOUDY WATERS

**1** *Fill the glass jar with water. Shine a flashlight through the water-filled jar onto a white screen behind it. You will see that white light from the flashlight travels all the way through the jar and the water.*

## SEE-THROUGH PAPER

*Hold a sheet of white paper up to the light, and it appears opaque. Now soak the paper in water, hold it up, and you will see that light can pass through it. This is because the water that soaks into the paper helps light pass through the gaps between the tiny paper fibers.*

**2** Pour a little milk into the water and stir it well to form a cloudy liquid. Shine the flashlight through the liquid again and this time you will see that the light on the screen is an orange color.

**3** Add more milk to the water until the liquid turns completely white. Shine the flashlight on the glass once more, and you will see that no light is able to travel through the glass and the liquid, leaving just a shadow of the glass on the screen.

## WHY IT WORKS

When you pour milk into the water, the milk particles that float in the liquid scatter some of the colors that make up the white light (see the splitting light experiment on pages 20-21), leaving only orange and red to travel through. As a result, the liquid is translucent. When you add more milk, the particles block out the light completely, making the liquid opaque.

ORANGE AND RED LIGHT PASS THROUGH THE CLOUDY LIQUID

ALL THE SPECTRUM'S COLORS ENTER THE GLASS

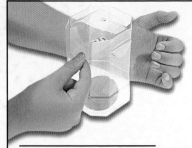

# BENDING LIGHT

THE EXPERIMENT ON PAGES 12-13 showed you how light can be reflected to turn corners, but it can also be helpful to bend light rays slightly. This bending is called refraction. Eyeglasses use refraction to bend rays of light, helping people to see things clearly. Microscopes also bend light rays to make objects appear bigger. Make a simple microscope in this project and see how bending light rays can be useful.

**WHAT YOU NEED**
*Clear plastic bottle*
*Small, flat mirror*
*Modeling clay*
*Drop of water*
*Scissors*

## MICROSCOPE

**2** *Ask an adult to cut two horizontal slits in the other two sides of the bottle, near the top.*

**1** *Ask an adult to cut the top off a plastic bottle and then cut out a narrow strip from two opposite sides of the lower half of the bottle.*

**3** *Push the two ends of one of the plastic strips through these two slits to form a platform.*

**4** *Place the mirror in the bottom of the bottle and angle it so it reflects light upward by propping it up with the modeling clay. Put something you want to look at close-up on the other plastic strip. Place a drop of water on the platform and look through the water at the object. You will see that the object appears larger, and you can study it in detail.*

## WHY IT WORKS

*The drop of water on the platform acts as a small lens. When light rays from the object pass through the drop of water, they are bent, or refracted. This occurs in such a way that they make the object appear larger when you look at it.*

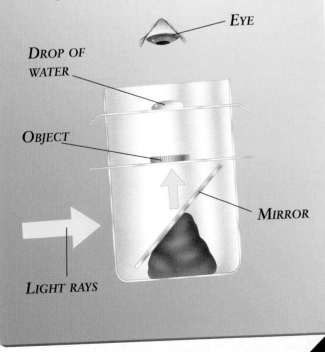

EYE

DROP OF WATER

OBJECT

MIRROR

LIGHT RAYS

## BROKEN PENCILS

*Fill a glass with water and place a pencil in it. When you look at the pencil from the side it appears as if the part of the pencil in the water has become bent or broken. This is because the water refracts, or bends, light rays from the pencil, making the underwater part appear as if it is in a slightly different position.*

17

# BRINGING IT CLOSER

TELESCOPES HAVE BEEN USED FOR HUNDREDS OF YEARS to look at faraway objects. Just like the microscope you made on the previous page, many telescopes use bulging, or convex, lenses to bend, or refract, light rays. This makes objects appear closer than they actually are. Build your own telescope in this project and see how it can boost your power of vision.

## WHAT YOU NEED
*One wide cardboard tube*
*One narrow cardboard tube*
*Two convex lenses*
*Eyepiece*
*Modeling clay*

## LOOKING INTO THE DISTANCE

**1** *Fit one of the convex lenses in one end of the wide cardboard tube.*

**2** *Fit the other lens into the eyepiece.*

**3** *Insert the eyepiece into one end of the narrow cardboard tube. Hold it in place using some of the modeling clay.*

## WHY IT WORKS

The lens at the front of the telescope gathers light rays from far away, while the lens in the eyepiece bends these light rays again to produce a larger image. By sliding the tubes back and forth you can make objects at different distances come into focus.

LIGHT RAYS

**4** Push the narrow tube into the wide tube, making sure they slide together smoothly.

## FOCUSING LIGHT

*Hold one of the lenses in front of a flashlight and shine the light onto a wall. Now move the lens toward and away from the flashlight. You will see that the size of the light beam changes as you move the lens. You should be able to focus the light into a small dot on the wall.*

**5** Hold your telescope up to your eye and slide the tubes into and out of each other until a distant object comes into focus and appears much closer.

# SPLITTING LIGHT

IN THE EXPERIMENT ON PAGES 14-15, you learned that light can come in different colors. This is because what we see as white light is actually lots of different colors mixed together. Sometimes these colors split to form a rainbow. This multicolored band of light is called a spectrum. You can split white light up into a spectrum with this simple project.

## WHAT YOU NEED
*Large glass bowl*
*Bulldog clips*
*Adhesive tape*
*Black cardboard*
*White cardboard*
*Mirror*
*Scissors*

## SPLITTING WHITE LIGHT

**1** *Seal the edges of the mirror with the adhesive tape.*

**2** *Ask an adult to cut a narrow slit in the middle of the black cardboard.*

**3** *Half fill the glass bowl with water.*

## COMPACT COLORS

*Look at the playing surface of a compact disc. The tiny notches etched onto the surface split light up,* *creating a spectrum that you will be able to see as you view the compact disc from different angles.*

**4** Using the bulldog clips, fix the mirror in the bowl of water so that it rests at an angle.

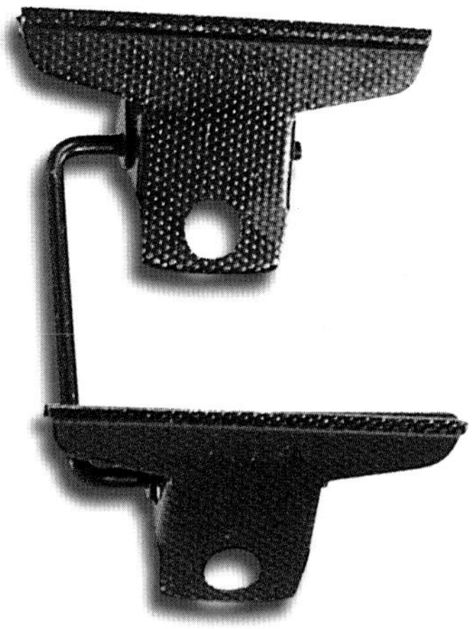

**5** Point the mirror at a bright window and place the black cardboard in front of it. Now place the white cardboard below the slit and adjust the mirror until you see a spectrum appear on the piece of white cardboard.

## WHY IT WORKS

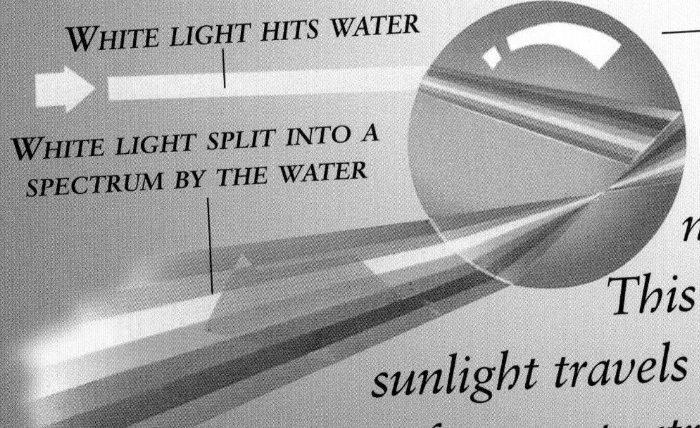

WHITE LIGHT HITS WATER

WHITE LIGHT SPLIT INTO A SPECTRUM BY THE WATER

The angled mirror creates a triangular region near the surface of the water. This shape is called a prism. As sunlight travels through the prism, the light is split up to form a spectrum. Raindrops act as tiny prisms, splitting sunlight to create rainbows.

21

# MIXING LIGHT

JUST AS YOU CAN SPLIT LIGHT INTO DIFFERENT COLORS (see pages 20-21), so you can mix colored lights together. Televisions use this principle to make color pictures. Look very closely at your television screen (not for too long, though!), and you will see that it is made up of thousands of tiny blue, red, and green dots. This project shows how you can mix colored lights to form new colors.

## WHAT YOU NEED
*Three flashlights*
*Red, blue, and green plastic wrap*
*Three cardboard tubes*
*Adhesive tape*
*White cardboard*

## COLOR WHEELS

*Color a circle of white cardboard with the colors of the spectrum in different segments. Ask an adult to push a sharp pencil through the middle. Then spin the wheel as fast as you can. The colors on the wheel will appear to blur and mix, making the wheel look white.*

## COLORED SPOTS

**1** *Cut out squares of the red, blue, and green plastic wrap that are big enough to fit over the ends of the cardboard tubes.*

**2** *Fix the plastic wrap to the ends of the tubes using the adhesive tape.*

## WHY IT WORKS

The new colors are made by mixing the three colored lights on the white background. For example, red and green will form yellow. If you mixed all three colors together, they would form white light!

**3** Place a sheet of white cardboard on the floor in a darkened room. With a couple of friends, hold the tubes and shine the flashlights down them onto the white cardboard.

**4** Now move the spots of colored light so that they overlap and create new colors of light on the cardboard.

# Separating colors

Mixing colored lights (see pages 22-23) is not the only way to create new colors. Just like the colors on a television screen, the colors on this page are made up of tiny dots. This time the colors of these dots are magenta (pink), cyan (blue), yellow, and black. These colors, called pigments, are mixed together in different quantities to create all the other colors.

### What you need
*Water-based felt-tip pens*
*Blotting paper*
*Bowl of water*

## SPLITTING PIGMENTS

**1** Cut the blotting paper into strips and draw patterns on them using a different colored marker for each strip.

**2** Place the strips in the bowl so that only their bottoms are in the water while the rest hangs over the side.

**3** Watch as the water soaks into the strips of blotting paper and starts to separate the ink into its different colored ingredients, or pigments.

**4** When the water has completely soaked the strips of blotting paper, take them out and closely examine the strips to see which pigments make up each color.

## INKY RINGS

Cut out a circle of blotting paper. Draw a large spot in the middle, using a water-based marker. Then cut a strip from the spot in the center to the edge and fold it down so that it hangs in the water. As the water soaks into the paper, the color will separate, forming circles of pigments.

As the water soaks into the strips of blotting paper, it carries the pigments with it because they are water soluble (they can mix with water). However, the different pigments that make up the colors are carried by the water at different rates. As a result, the pigments are separated into bands, allowing you to see which pigments make up the colors. For example, red ink is made up of yellow and magenta pigments.

# LIVING PICTURES

ALL THROUGH THIS BOOK YOU HAVE SEEN the different aspects of light, from what light is made up of to how light rays can be altered to enlarge or color things. One other practical use for light is to create moving images, such as the ones you see when you go to the movies. This experiment will show you how moving pictures "move" and how this is all an illusion!

## WHAT YOU NEED
*Cardboard*
*Colored pens*
*Nail*
*Bead*
*Cork*
*Mirror*

## FLYING PARROT

**2** *Color the pictures using the colored pens. Ask an adult to cut out a series of slits around the middle of the circle using a sharp knife.*

**1** *Ask an adult to cut out a large circle of cardboard. Then draw the images in the design shown on the opposite page.*

**3** *Ask an adult to pierce the center of the cardboard circle with the nail. Slide the bead onto the nail and then push the cork on to make a handle.*

## FLIP BOOK

*Draw each of the parrot illustrations on the corner of a pad of paper. Now flip through the pages very quickly using your thumb. Once again, the parrot appears to move.*

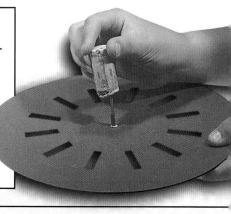

# WHY IT WORKS

You are not really seeing a moving parrot, but a series of slightly different pictures. When these pictures are seen very quickly, one after the other, they give the impression that the parrot is moving. Movie projectors work in the same way. They flash separate still images onto a screen very quickly, making it look like the picture is moving — when really it isn't!

**4** Ask a friend to hold the mirror in front of you. Look through the slits from the back of your cardboard disk and spin it quickly. You will see that the parrot looks as if it is flying.

# BEAMS OF LIGHT

As well as projecting moving pictures onto a movie screen (see pages 26-27), light can be used to carry lots of information along miles and miles (km) of special glass wires called fiber-optic cables. These beams of light can carry sound and pictures, such as telephone conversations and television pictures. This experiment shows you how these cables can carry light rays over the most winding routes, despite the fact that light travels in a straight line.

**WHAT YOU NEED**
*Black paint*
*Flashlight*
*Pin*
*Large glass bowl*
*Clear plastic bottle*

## FIBER OPTICS

**1** *Ask an adult to cut the top off the plastic bottle. Paint the outside of the bottle black, leaving a small area clear on one side. Using the pin, make a small hole in the bottle on the opposite side from the clear area.*

## REMOTE CONTROL

*Your TV remote control uses invisible beams of light to change channels. Try covering the front of the remote control with your hand. You will find that the remote control will not work because your hand blocks the light.*

**2** In a darkened room, stand the bottle against one side of the bowl so that the clear area faces out. Fill the bottle with water and shine the flashlight through the clear area. Place your finger in the stream of water and you should be able to see a spot of light.

## WHY IT WORKS

The stream of water acts like a fiber-optic wire. As the rays of light travel down the stream, they bounce, or reflect, off the sides, traveling along the stream even as it bends. The rays of light hit the stream's sides at such a shallow angle that they are reflected inside. In the same way, fiber-optic cables carry light rays along a curved path by reflecting them off the sides of the fibers.

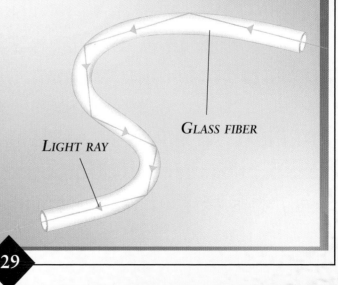

GLASS FIBER

LIGHT RAY

# GLOSSARY

**CARBON DIOXIDE** (KAR-bin dy-OK-syd)  A gas that plants take in from the air and use to make food. *Find out about how plants use carbon dioxide, sunlight, and water to make their food on page 7.*

**MIRRORS** (MIR-urz)  Flat surfaces that show an exact picture of something placed in front of them. *See mirrors in action reflecting light on pages 12-13.*

**OPAQUE** (oh-PAYK)  When no light can pass through an object, it is called opaque. *Find out how to make water opaque on pages 14-15, and see the effects of opaque objects with your shadow theater on pages 10-11.*

**PERISCOPE** (PER-uh-skohp)  A tool that is used to see above the surface of the water from below the surface. *Build your own periscope on pages 12-13.*

**REFLECTION** (rih-FLEK-shun)  This occurs when light rays bounce off a shiny surface, such as a mirror. *Turn to pages 12-13 and see how reflection makes a periscope work. Can you find any other examples of reflection. Hint: look on pages 28-29.*

**REFRACTION** (rih-FRAK-shun)  This occurs when light rays are bent as they travel through an object. *See refraction in action on pages 16-17 and 18-19 when you build a microscope and a telescope.*

*SHADOW* (SHA-doh)  A dark area caused by an object blocking out light rays. *Build your own shadow theater on pages 10-11.*

*SPECTRUM* (SPEK-trum)  Using a specially shaped piece of glass called a prism, sunlight can be split up into a band of colors. This band is called a spectrum. *You can see a spectrum on pages 20-21.*

*TELESCOPE* (TEH-leh-skohp)  An instrument used to make distant objects appear closer and larger. *You can find out how to build a telescope on pages 18-19.*

*TRANSLUCENT* (trants-LOO-sent)  When only some light rays are able to pass through an object, it is called translucent. *See how you can make water translucent by adding a small amount of milk on pages 14-15.*

*TRANSPARENT* (tranz-PER-ent)  Objects are transparent when they let light pass through them freely. As a result, you can see through them completely. *Take a look through the book and find some transparent objects. Then see if you can find any more transparent objects around your home.*

# INDEX

cameras  8, 9

colors  15, 20, 21, 22, 23, 24, 25, 31

compact discs  21

eyes  9, 13, 17, 19

fiber optics  28, 29

focus  18, 19

lenses  17, 18, 19

light rays  8, 9, 10, 12, 13, 16, 17, 18, 26, 28, 29, 30, 31

microscopes  16-17, 18, 30

mirrors  12, 13, 16, 17, 20, 21, 30

movies  26, 27, 28

opaque  14, 15, 30

penumbras  11

periscopes  12, 13, 30

pictures  22, 26, 27, 28

pigments  24, 25

plants  6, 7

pupils  9

puppets  10-11

rainbows  20

reflection  12, 13, 14, 16, 17, 28, 29, 30

refraction  16, 17, 18, 30

retinas  9

shadows  8, 10, 11, 15, 30

spectrums  15, 20, 21, 22, 31

sun, the  6, 7, 8, 31

telescopes  18, 19, 30

televisions  22, 24, 28, 29

translucent  14, 15, 31

transparent 14, 31

umbras  10

water  6, 7, 14, 15, 16, 17, 20, 21, 24, 25, 28, 29